DEBUNKING
GENERAL RELATIVITY

An Understanding of Dark Matter
That Shows Newtonian Theory Was Correct

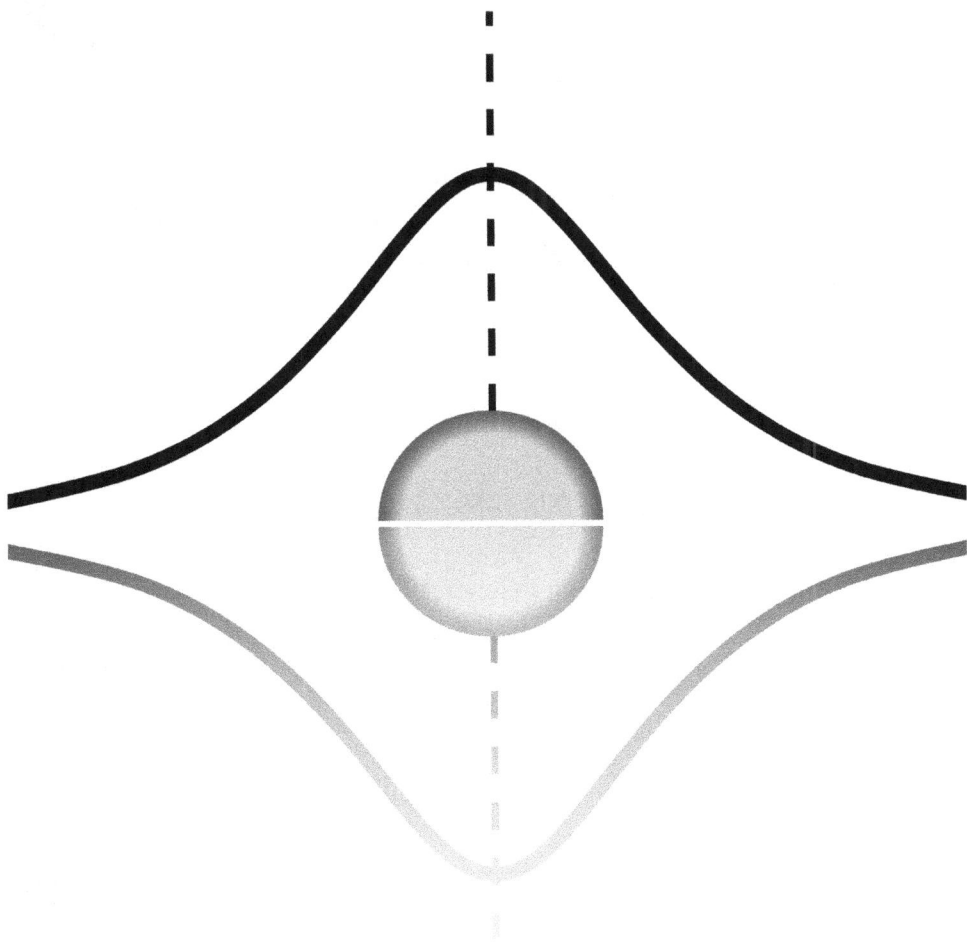

GARY LEDERER

Printed in the United States of America

First Printing, 2016

ISBN 978-0-692-71958-9

www.DebunkingGeneralRelativity.com

Dedicated to my grandchildren.

DEBUNKING GENERAL RELATIVITY

This book is an analysis of gravitational attraction from someone possessing a talent to reverse-engineer complex processes and correctly analyze how they function. I believe there is adequate information available to produce a detailed explanation of how gravity works, with no preconceived limitations. Over time, research will turn in the direction of this concept and find it to be correct. Here I reexamine existing observations and theories of gravitational attraction, taking a logical look at gravity while continually trying to ask the right questions and inquire, "How does it do that?" I review problems and conflicts with existing theories, which allows the reverse engineering of gravitational force and dark matter. I present a hypothesis concerning the properties of dark matter and its contribution to gravitational force.

Here is a list of problems and observational conflicts with current gravitational theories.

1. General relativity breaks down at the molecular level.

2. There are unsolved problems about the outer reaches of our solar system from 100 years ago that need to be reexamined.

    a. Einstein understood there was a problem with the concept of a linear force traveling faster than the speed of light to instantaneously reach from the Sun to Neptune and beyond.

    b. His theory of general relativity replaced the linear force concept with the curvature of space-time whose influence is in place waiting for mass to arrive. This understanding has been stretched to compensate for the motion of the solar system around the galaxy and the galaxy through the universe.

c.   We now know there is an unknown, unseen dark matter that seems to contribute approximately 80% of the gravitational attraction within galaxies. I contend that this is an outside force that negates the assertion Einstein made about space being completely empty. This unknown force requires us to reexamine everything to determine if all or some of general relativity is obsolete. Dark matter causes a discrepancy in the orbital speeds of stars within center discs of galaxies when compared to the orbits of planets in our solar system. The speed of these stars is in direct conflict with general relativity.

3.   The conflict of mass or information being lost from our universe into a black hole is an impossible consequence for quantum theory.

4.   We must look at Mercury's orbit to review the origins of general relativity, and then present alternate explanations for lensing and variations in time.

It has been 100 years and no one can explain how mass actually curves space-time. How does it do that? It seems to be necessary to review the information that was known 100 years ago to remind everyone of general relativity's original parameters. This list compares what was known in 1915 to observations in 2016: In 1915, the Milky Way was thought to be the entire universe. It's now known to be one of billions of moving galaxies.

1.   Astronomers suspected our solar system was moving in space but it was not known for sure. Now we understand all galaxies are rotating and moving through space.

2.   In 1915, our solar system was only known out to Neptune, 2.8 billion miles from the Sun (except the occasional comet). Today it is

known to extend past the Kuiper belt and Oort cloud, which extend to 5 billion miles and more than 10 billion miles respectively.

3. Planets beyond Mars were referred to as gas giants. This is because the calculations for gravitational attraction could not account for a massive object traveling at the speeds observed. Uranus and Neptune are ice-covered planets now estimated to be 14.5 and 17 times the mass of Earth respectively. These planets are far more massive than they were thought to be 100 years ago.

4. Albert Einstein assumed that outer space was an empty vacuum and the planets and other objects moved unobstructed through it. We now accept that space around galaxies is occupied by dark matter along with other probable outside influences on orbits.

5. It took three and a half years to achieve photographic proof of the light curvature predicted by general relativity. Approximately four years after that Georges Lemaitre, followed by Edwin Hubble, discovered the existence of other distant rotating galaxies. The next conclusion was that the Milky Way was one of many rotating galaxies. This doubled down on the fact that the vacuum of outer space must not be in any way obstructing the movement of mass. This is the only scenario that could allow mass' momentum to carry it around the curvature of space while the curvature was moving. This prerequisite is essential for the entire solar system to keep pace with the Sun moving around the galaxy. These discoveries were accompanied by calculations by Jacobus Kapteyn and Jan Oort that led to the need for dark matter's additional attraction.

6. There are at least three outside influences known today to be acting upon mass in outer space that were not considered 100 years ago. It is time to account for them.

a. An assortment of microscopic particles within the supersonic solar wind is moving outward from the stars continuously. This solar wind certainly has some interaction and influence on orbits.

b. If there is any interaction of magnetic fields in space, it would constitute an impediment to a mass's orbit. We now know the Sun's magnetic field extends out past Neptune. That means Earth's magnetic field is completely engulfed within the Sun's magnetic field and surely has some interaction.

c. The largest interference with mass's motion is the dark matter that is everywhere within galaxies and is estimated to be responsible for 80% of the attraction known today.

7. In the 1930s, it became obvious that general relativity broke down at the molecular level. This initiated the origin of quantum theory and the graviton.

CONCLUSION N0.1

When calculations are completed to account for the objects over 10 billion miles from the Sun, I am confident that the current formula for general relativity will not be sufficient. It will take the additional attraction that can only be attributed to dark matter to make it function correctly at these distances, just as dark matter was needed to explain the motion of stars within galaxies' discs. The addition of dark matter's attraction is also needed to form galaxies and hold them together in computer models of galaxies.

CONCLUSION N0.2

It is time to conclude that mass is traveling through space within a medium of dark matter. Moreover, dark matter and observations of it are in direct conflict with general relativity's prerequisite. This is why dark matter's presence is so puzzling to today's scientific community. My theory boils down to one basic disagreement: Is mass traveling unobstructed through the vacuum of empty space as general relativity predicates, or is dark matter the medium that mass is inside of moving through space? If there is a medium, general relativity's analogy must be discarded, and this means the interaction of mass with dark matter must and does mimic the curvature of space-time in the inner portion of our solar system. It not only produces all the characteristics explained by general relativity but it also solves all other unexplained problems that have been observed in galaxies' discs, the outer solar system and also down to the microscopic level. This book shows how the formulas of general relativity work within this scenario and reinforce the formula $E = mc^2$.

HYPOTHESIS

Of Dark Matter's Characteristics

After I determined that general relativity had been stretched far beyond its original parameters and must be discarded due to the presence of dark matter, I searched for some characteristics of dark matter that could produce the observed actions. Some radical conclusions came to light about dark matter when I used known observations to reverse-engineer gravitational force.

1. Dark matter not only produces gravitational force; it also conducts this force.

2. Time has shown that matter is engulfed within a medium of moving dark matter that is following the dominant mass at the center of the galaxy.

3. Mass alters the density and polarizes this reservoir of dark matter toward the dominant mass in the vicinity.

4. Any build-up of dark matter in front of a moving mass acts as an attraction increasing motion instead of acting as a resistance impeding motion.

5. There is a quantity of dark matter which is proportional to the amount of gravitons within matter that naturally adheres to each atom.

6.   Dark matter is only one component of matter, which is the graviton. Dark matter should be referred to as free gravitons because it possesses attributes similar to gravitons but to a lesser degree due to its being less dense.

These conclusions present themselves as the mirror image of general relativity and expose a variable that expands and enhances its coverage. Let's stop ignoring the circumstances under which the theory and its mandates originated. We now know the mandate about empty space is not true since we accept dark matter's presence.

There have been a series of observations in conflict with general relativity spread over nearly 90 years, but the original understanding of this theory has been stretched to encompass these problems. The formulas do work in our inner solar system; however, the main reason to stretch general relativity is because there has been no other tangible reason for the curvature of light near massive objects. You can touch the third rail and say, "There might be a problem with general relativity." It takes a blending of multiple theories to understand how gravitational attraction works.

DARK MATTER

The confirmation of dark matter's existence by Vera Rubin that showed how stars moved within the center disc of galaxies made it clear that there was a distinct difference in how stars move compared to how planets move. This in turn redirected my focus on how gravitational force is additive within mass. I then determined that dark matter affects every aspect of gravitational force from the molecular level up to the largest stars. To understand dark matter, you must be willing to question general relativity.

The particle that is within matter, creating the attraction to other matter, (the elusive graviton) is conducting the attractive force it receives from adjoining mass and adding its own contribution as it passes it along. After spending a career working with electrical circuits and voltage drop, it is apparent to me that within mass the density of gravitons has the characteristics of a series circuit passing the force along and adding its contribution to the attraction. This is possible because the particles are densely packed within matter and the mass itself is tightly packed. The term dark matter is incorrect because it is only one component of matter. It can never accumulate into matter because it is lacking neutrons, protons and electrons. These particles in space are known as dark matter, but they should be referred to as free gravitons. Dark matter and free gravitons are synonymous in any example from here on.

It is clear that the particles in space surrounding mass within our solar system are only dense enough to pass this force along at the rate of decrease that has been discovered to be approximately the inverse of distance squared. The free gravitons surrounding our solar system are poor conductors because of their low density. It is not clear that the base free graviton density is consistent throughout all galaxies. It seems as though the locations within galaxies that contain a high concentration of stars has resulted in an increase in the base density of free gravitons within that section of the reservoir. This increases the overall conductive ability in these areas.

There is a minimum attraction within galaxies missing in today's understanding of gravitational force along with the buildup of free gravitons caused by masses' accumulation. The base density plus the buildup of polarized free gravitons create the attraction instead of a geometric shape of space-time. This combination allows the attractive force of our Sun to extend much further than previously calculated which accounts for the items as far out as the Oort cloud to stay in orbit while moving through space. This also solves multiple mysteries in motion that are being discovered within galaxies.

There is a reservoir of free gravitons that each galaxy is contained within. The movement of the reservoir of free gravitons extends the attraction and motion from the super massive object at a galaxy's center all the way to the outer reaches of each galaxy. This object at the center of galaxies is not a black hole; it is a gravitational dark star (GDS). The free gravitons surrounding this object have reached a density that completely blocks light. At the point this occurs is referred to as the event horizon. The density of free gravitons affects change and motion which is how we measure time. Time does not stop at the event horizon as previously calculated, although our perception of time is altered by the extreme limitations of change and motion.

PARTICLES WITHIN MATTER

Within matter, science has identified a great deal about the nucleus of an atom including breaking the neutrons and protons down into much smaller parts. The atomic model of the electrons has come into question recently, but the size of the nucleus compared to the overall size of an atom remains miniscule. If the nucleus is scaled up to be represented as the size of a tennis ball, the atom's overall area would become the size of a sports stadium. The consensus of this analogy is that the space around the ball   is empty except for electrons, yet no light passes through the stadium.

I contend this area surrounding the nucleus is filled with a high density of what is thought of as dark matter. This area is the location of gravitons. In the periodic table, there are very few elements that are transparent and even those that are transparent, such as the gases, deflect some light. The majority of the elements contain a density of gravitons surrounding their nuclei that block light. Each element absorbs or reflects a unique spectrum of light due to this internal density of gravitons.

An atom of matter within an area of free gravitons appears to have the ability to attract and control a small amount of those free gravitons. The density of gravitons within each element determines the proportional amount of free gravitons that cling to that atom. Every atom reaches a natural balance of gravitons inside and free gravitons outside. This proportion is consistent and is the fundamental basis for gravitational attraction.

The free gravitons increase the area of attractional influence for each atom. As mass accumulates, the internal pressure reduces the space available for the free gravitons and they are forced to migrate outside the mass into the reservoir. This upsets the natural balance and the free gravitons are then held close to the mass increasing the density of the reservoir around it. This polarizes the displaced free gravitons and the reservoir itself back towards the mass the free gravitons were forced away from. This polarization appears to continue outward through the moving reservoir many times further than the increased density reaches. The internal pressure of mass also appears to initiate a rotation to objects once they have accumulated to a point when the mass forms itself into a sphere. This is the difference between an object tumbling through space and an object rotating on its axis. When two objects each attract the same free gravitons, this constitutes an attraction between those two objects.

Gravitons inside mass and also surrounding it resembles the Higgs' Field, only the Higgs' particle is thought to be a weak nuclear force boson. My theory assumes that gravitons and free gravitons require the particle to have the characteristics of a strong nuclear force gluon. Gluon pairs being responsible for gravitational attraction is apparently fundamental to other theories of gravitational attraction. Quantum theory has recently altered things with the uncertain distinction between particles and waves; therefore, whether the graviton is a string or particle functioning as a wave is a distinction that I cannot make definitively.

Free gravitons are not visible, yet they fill every void around atoms within earth's atmosphere. At sea level atmospheric pressure forces atoms together, reducing the space available for free gravitons when compared to the upper atmosphere. Downward into the planet, the pressure continues to increase and the atoms pack more densely closer to the center of the planet. This creates denser and denser mass that increases the generation and conductivity of gravitational force because gravitons within this mass are denser than the free gravitons that have been forced to migrate outside the mass.

Down at the molecular level, general relativity breaks down, but my inverted view of its formula correctly represents the variations in free graviton density that I refer to as being fundamental to gravitational attraction. The internal pressure of masses' accumulation is the key to gravitational fields because it disrupts the balance of free gravitons that naturally cling to each atom of matter.

When gravitons are forcibly released from within matter, their density immediately dilutes into the reservoir surrounding it. This is why discovering them is so difficult and why it appears this space within matter is as empty as the area outside the atom. My view of the entire system is more representative of the formula $E = mc^2$ than is the curvature of space-time, and my theory shows a clear connection between gravitational force and special relativity that I feel Einstein was moving toward. His concept was the mirror image of the true description because he was missing the contribution of the gravitons and free gravitons. It is more logical that a tiny particle can affect other similar particles than that a tiny particle could curve space-time. It is easy to see how some theorists contend this particle or string exists in another dimension because it is so difficult to actually see. It is only apparent when it is dense enough to affect light.

When the internal pressure of mass gets high enough the atoms themselves fuse together. This fusion within stars consumes free gravitons or expels more depending on the elements being fused and produced. When the accumulation of free gravitons surrounding a star becomes as dense as gravitons inside most matter, light is blocked completely. This produces a gravitational dark star at its event horizon. A collapsing star can quickly force the transition from a brilliant star to a gravitational dark star when the outward force of fusion diminishes and the internal pressure skyrockets. When a super nova erupts, all the matter released and/or produced reclaims their allotment of free gravitons that pressure had forced to migrate outside that star.

This collection of free gravitons surrounding a mass and blocking all light resembles the gravitons that were previously discussed blocking the light in a scaled-up, stadium-sized atom. There is so much internal pressure on the mass that gravitons are being forced outside the atoms, again blocking light. This deficit in the natural balance of gravitons and free gravitons

16

within super massive gravitational dark stars causes a polarization in the reservoir to extend to the outer reaches of the galaxy. Then as the gravitational dark star rotates and moves through the universe the entire reservoir of free gravitons follows.

The free gravitons surrounding this massive object do not constitute complete matter because they have no nuclei or electrons. The matter inside the gravitational dark star most likely has a deficit of gravitons within its atoms so most of these atoms no longer resemble matter as we would recognize it. There is no information lost from the universe because there is no hole in space. Quantum theory has been in conflict with losing information into a hole in space for so long that the waterfall type depiction of a black hole has transitioned into a large dense black object in most depictions. The gravitational dark star surely releases some electromagnetic force under certain circumstances as it consumes matter. It appears that the electromagnetic forces can escape along the mass's magnetic force lines.

Apparently there is a gravitational slip that occurs at the event horizon because most gravitational dark stars, quasars, etc. are rotating extremely fast. If there was no slip somewhere, the outer portion of the galaxy would be moving so fast it would fly off into space. I conclude that all rotating objects with their own gravitational field have some degree of "slip" when compared to a nonrotating object of identical mass. In other words, all motion has some effect on gravitational attraction even though it might be minute.

NEPTUNE'S ORBIT

Since I am discarding the analogy of general relativity, which was hypothesized to be in place ahead of an orbit to overcome the speed of light problem, we must review Neptune's orbit to present an alternative. This alternative theory not only duplicates the aspect of being in place ahead of an orbit but it clarifies the difference in the movement of stars compared to planets and adds the attraction of free gravitons to all gravitational calculations. Let's look at what scientists recognized as the outer reaches of our solar system, where Neptune is approx. 2.8 billion miles from the Sun and its orbit takes approximately 165 Earth years.

One hundred years ago, Einstein observed a problem with gravitational attractions at these distances and further. The problem was if gravitational force is linear it has to travel faster than the speed of light to attract an object to the correct point in space at a given moment. Einstein determined that nothing could travel faster than the speed of light. His development of general relativity fixed the linear attraction problems by saying that mass curves space-time and planets move in a straight line that follows the curvature of space-time established by the Sun. The curved space is there waiting for the planet to arrive so there is no conflict with attraction traveling faster than the speed of light.

Einstein's theory was predicated on an assumption that outer space is an empty vacuum that does not impede the movement of objects. Clearly, it was not meant to compensate for the movement of the Sun and our solar system around the galaxy because in 1915 scientists were not sure if the Sun was stationary or moving through space. In the 1920s, the discovery of separate galaxies by Georges Lemaitre and Edwin Hubble required a change in thought about the motion of our solar system and planets. This was so soon after the 1919 photographic confirmation of light curvature during an eclipse that the scientific community accepted this change.

Under the 1915 observations, the orbit of a planet like Neptune would have been nearly consistent at 12,000 mph around the Sun and through space. Under the change made in the 1920's, the exact speed was not clear, but the understanding of Neptune's orbit now is that Neptune travels through space at speeds varying from approx. 488,000 mph when traveling opposite the Sun's motion to 512,000 mph when traveling with the Sun's motion. These calculations of Neptune traveling through space do not account for the significant motion of the Milky Way through the Universe. This variation is even greater with planets closer to the Sun. This is a huge change to the understanding of a planet's orbit but was accepted to be compatible with general relativity because of the 1919 confirmation of light curvature and there being no other explanation for that phenomenon. This variation in speeds is now accepted as a fact even though there is clear evidence that space is not empty and there are outside influences on orbits.

The outer reaches of the Oort cloud are many times further from the Sun than Neptune and the orbital motion of these objects appears negligible around the Sun yet they are moving through space at approx. 500,000 mph. The calculations to keep these distant objects in unison require a new understanding of attraction and also a distinction between the two types of motion - motion through space and the motion of an orbit around a star.

SOLUTION TO NEPTUNE'S ORBIT

(Plus Other More Distant Objects)

DIAGRAM NO.1

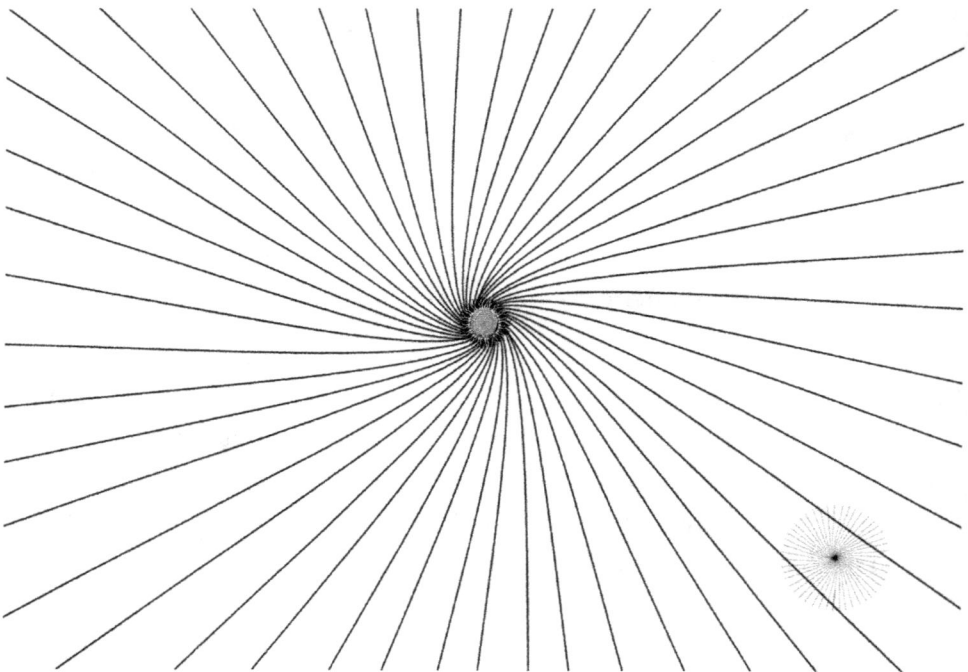

Diagram No. 1 depicts a rotating mass that has become the dominate mass in the vicinity which is our Sun. The smaller object represents a mass typical of a planet orbiting our Sun each controlling a proportional amount of free gravitons.

This example of our solar system moving through space within a conductive reservoir of free gravitons solves all of these problems. The free graviton reservoir is moving through space and a localized portion is polarized towards the dominant mass in the vicinity. The dominant mass also increases the density of the free gravitons closer to its mass. Like relativity this force's influence is there waiting for the planets to arrive due to the reservoir of free gravitons. Unlike relativity, where the curvature of space-time moves along with the Sun only, my model of free gravitons explains and clarifies how the Sun and its entire far-reaching solar system are moved along in unison with the reservoir of free gravitons.

Neptune is orbiting the Sun and is passing through the polarized reservoir of free gravitons around the Sun. Neptune has a much smaller polarized field of its own that is interacting with the Sun's field as it orbits. The reservoir of free gravitons is moving around the Milky Way and is bringing the Sun and subsequently its entire solar system along with it. This constitutes two distinctly different types of motion. It is the difference between moving along with free gravitons and moving through them. The Sun as the dominant mass has taken over a large area of the moving reservoir and polarized it, thus the Sun moves along with the reservoir. The planets have much smaller areas of influence and are traveling through the Sun's polarized portion of the reservoir.

It is logical that the free graviton particles in space that are polarized back toward an object follow the rotation of that dominant mass. This causes a curvature in the adjacent particles. This curvature is responsible for the lensing seen near large masses and is important because it is an accurate alternative for the lensing predicted in general relativity.

The dust cloud from which our solar system developed was not standing still. The cloud was moving through space within the reservoir of free gravitons. The matter in the cloud commandeered a proportionally predetermined amount of free gravitons to bring the random collection of matter together. At this point the entire cloud of matter was traveling through space around the galaxy at the same speed as the free graviton reservoir. The largest pieces of mass would have an advantage as this process began. As the cloud contracted around the dominant mass, the conservation of angular momentum influenced the entire cloud to rotate as it collapsed inward. This material moving around the dominant mass developed orbits traveling through the reservoir all the while still moving through space within the reservoir.

BALL PIT

When thinking of objects in our solar system moving through the free gravitons polarized by the Sun, I consider it to be the inverse of walking through a children's play area ball pit. Under normal circumstances, the slight increase in balls in front of you during motion would be thought of as a resistance; but with the reservoir of free gravitons, the increase in front of you creates a greater attraction that influences your forward motion. Also the size of an object determines the amount of free gravitons that build up in front of that object during motion.

The free gravitons with consistent density are functioning in a fashion that gives the result similar to what Albert Einstein viewed as mass moving through space unobstructed. As a mass travels into an area of higher free graviton density, this mass acts as if it was moving down an incline due to the increased attractive influence in front of it causing it to speed up. Transitioning into a less dense area lessens the buildup of attraction in front of the mass, and it slows as if it were moving up an incline. The free gravitons' force has the ability to compensate for other outside influences that are affecting mass during its orbit, such as solar wind and magnetic field interactions. There may also be other outside influences that have not been detected because the force of free gravitons is also overcoming their effects.

THE IMPORTANCE OF TWO TYPES OF MOTION

The previous section where Neptune's orbital motion was compared to the Sun's motion shows the importance in understanding the difference between the motion of stars and virtually all other objects within galaxies. Stars and near-star objects are certainly the dominant mass in most if not all individual areas of space within galaxies. Stars are "hitching" a ride within the moving reservoir of free gravitons by virtue of their dominance of an area. The speed of the stars is controlled by the speed of the rotating reservoir of free gravitons which is in turn controlled by the super massive gravitational dark star at the center of the galaxy. Binary stars move through each other's fields but share the dominance and follow the reservoir. These stars have established their dominance by consuming mass then controlling its associated free gravitons which also extend the polarizing attraction through the reservoir. Each of these is proportional to the total mass. This constitutes the first type of motion.

Apparently the overall amount of attraction that is credited to dark matter has been overestimated due to today's theory of gravitational attraction which does not take into account that the motion of the stars nearly matches the motion of the reservoir. Over time the build-up in momentum of the reservoir of free gravitons and all the mass that is weaved into it is skewing the calculations currently being attributed to its attractive contribution. The amount of attraction originating from free gravitons is probably closer to 50% than 80% of all gravitational attraction within galaxies.

The second type of motion observed involves all the smaller objects such as planets, etc. that have avoided being consumed by the star. They are sent traveling through the polarized reservoir. The largest remaining pieces build their own small increased density and polarized field that is also proportional to their mass. The interaction of these fields establishes the speed and orbit of the planets. This second type of motion is what Einstein knew of in the early 1900s. This is why he did not account for the stars' movement around the galaxy. There was not and still is not any tolerance for an alternate type of motion in gravitational theory. This is why dark matters' presence and additional attraction are not understood.

DIAGRAM NO. 2

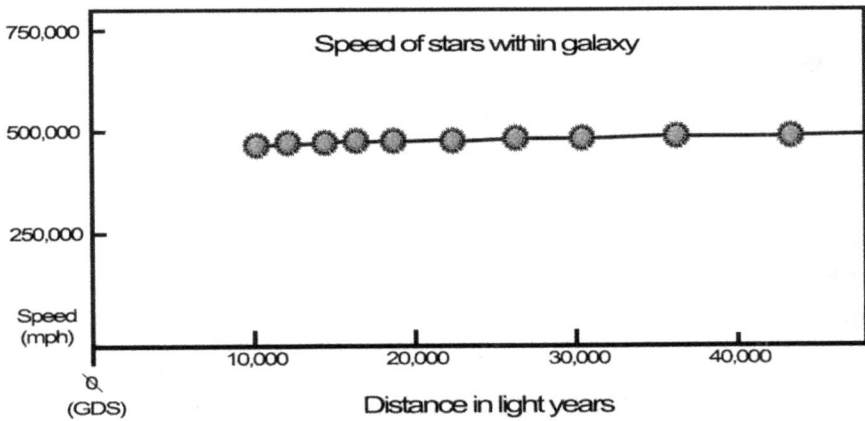

Diagram No. 2 depicts the observed motion of stars within a typical spiral galaxy that nearly matches the motion of the reservoir of free gravitons. These observed orbital movements are the confirmation of the existence of dark matter because they differ greatly compared to the orbital speeds of planets in our solar system and are in direct conflict with general relativity.

DIAGRAM NO. 3

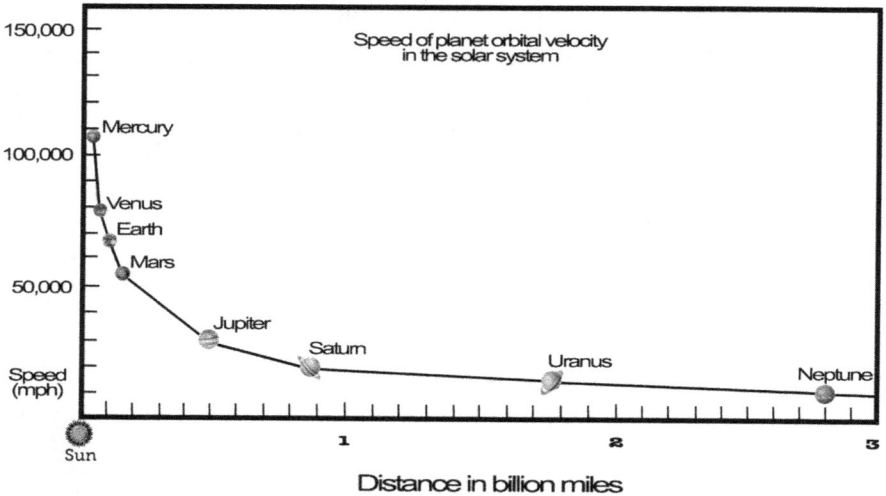

Speed of planet orbital velocity in the solar system

150,000

Mercury

100,000

Venus
Earth
Mars

50,000

Jupiter

Saturn

Uranus

Neptune

Speed (mph)

Sun

1

2

3

Distance in billion miles

Diagram No. 3 depicts the observed orbital speed of the planets within our solar system. The decline of orbital velocity over distance is completely different than the actions of stars around galaxies. The concept of curvature in space-time that was developed to match the planets' movements is incompatible with the observed motion of stars shown in Diagram No. 2. These two diagrams clarify the difference in the two types of motion .

When stars migrate into close proximity of the super-massive gravitational dark star at the center of a galaxy, they begin moving through the free gravitons surrounding this massive object. At this point, the stars are no longer the dominant mass in the vicinity so they respond to the super massive center as planets respond to stars elsewhere. If these stars move closer to the event horizon, they fly apart due to tidal influences. The difference between the stars' motions and the planets' motions is a problem that scientists continue to ignore by reaffirming general relativity.

26

MIRROR IMAGE OF
GENERAL RELATIVITY

DIAGRAM NO. 4

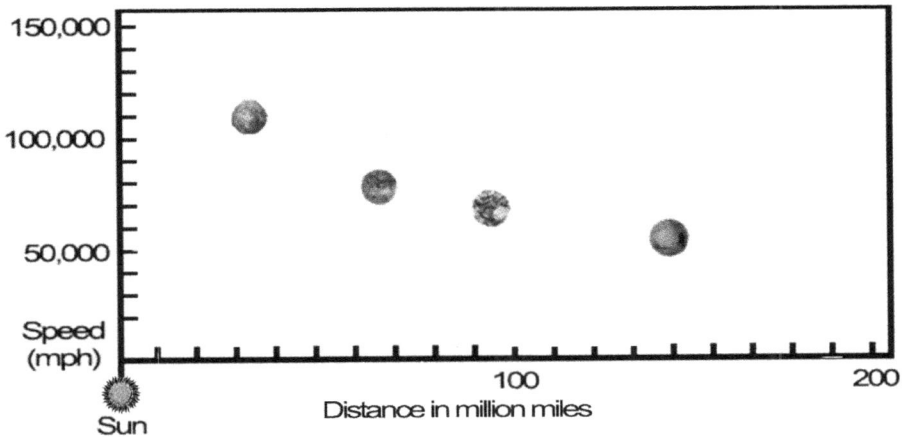

Diagram No. 4 depicts the orbital velocity of the four inner planets. It is clear that this information was the basis for general relativity. Connecting the dots on this graph reveals a curve that developed which is shown in Diagram No. 5. General relativity is the inverse of this curve.

DIAGRAM NO. 5

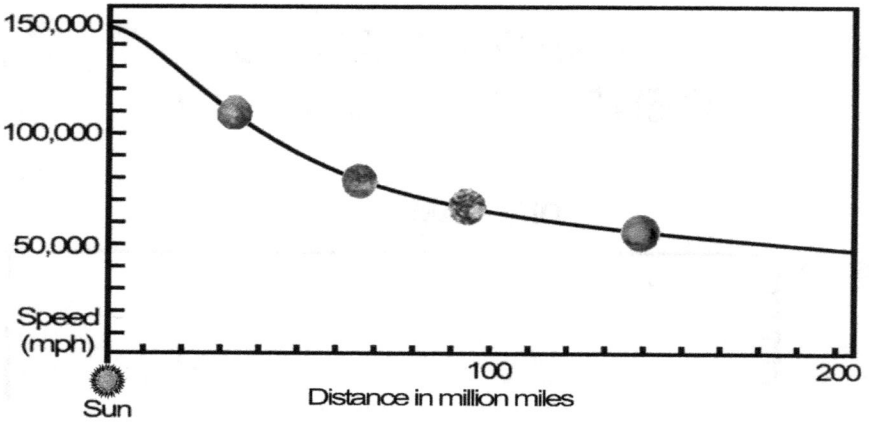

DIAGRAM NO. 6

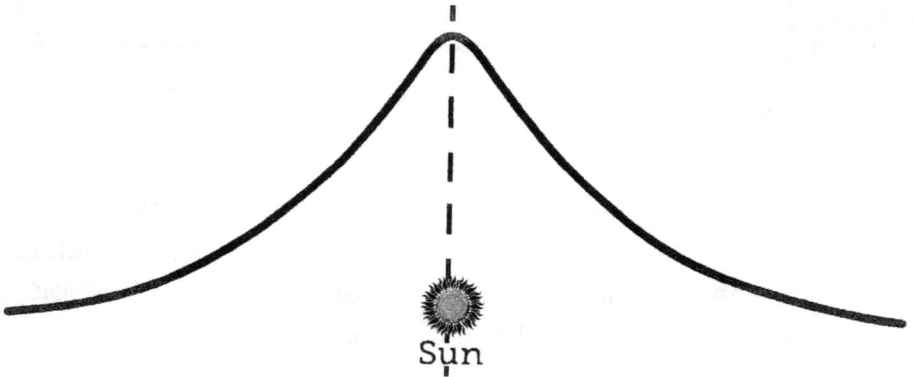

Diagram No. 6 depicts the curve derived from the four inner planets' speeds extended to the far side of the Sun. Diagram No. 7 shows the inverted version of Diagram No. 6 that is the beginning of the curvature of space.

DIAGRAM NO. 7

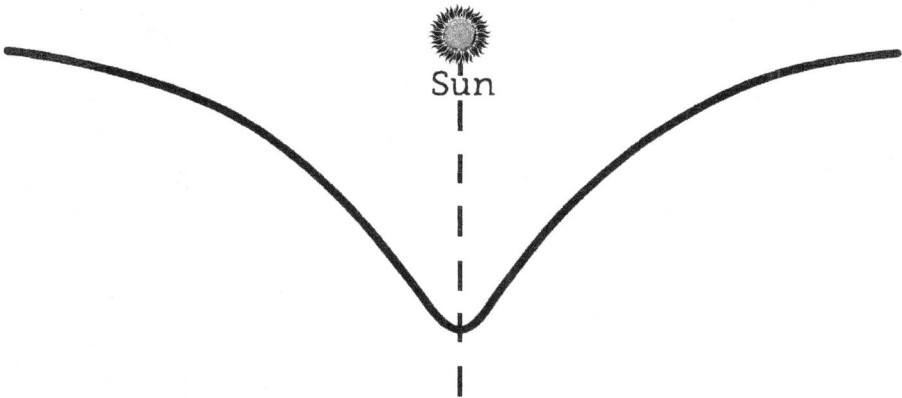

Sun

DIAGRAM NO. 8

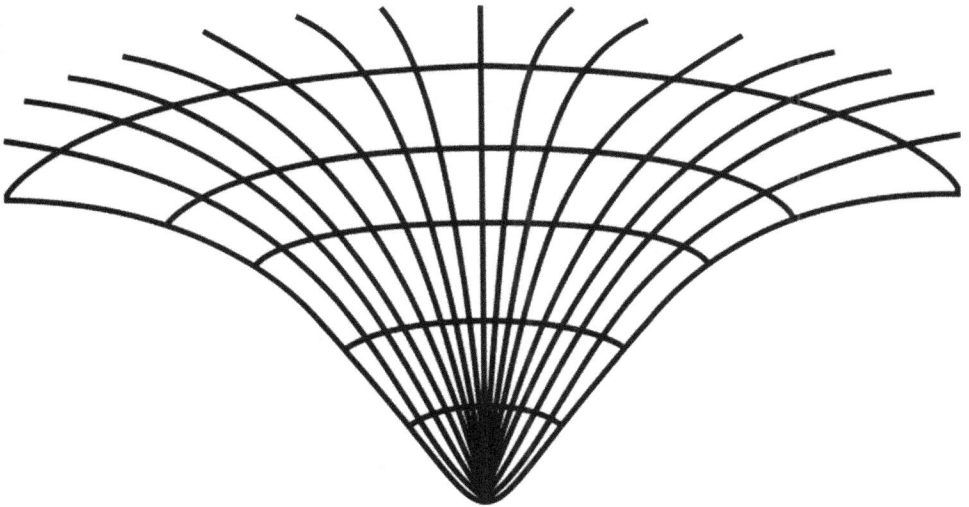

Diagram No. 8 depicts a cross section of the three-dimensional curvature of space. The bottom of the curve was derived from Diagram No. 7's inverted curve. The addition of time made this into a four-dimensional representation of a unified space-time curvature.

The geometric formulas presented by general relativity represent this information. This explanation is an over-simplification of a very complex process that went into the theory of general relativity.

Diagrams 4-8 show why general relativity's formula is accurate for objects orbiting a dominant mass because it was derived from a graph of the planets' actual orbits; however, it clearly falls short when applied at the molecular level or to the motion of stars. It has been 100 years since relativity was published, and no one can say how mass actually curves space-time. My theory of how gravity works also starts with the information that was shown in Diagram No. 5 where we connected the dots on the graph of the Inner Planets' Speeds. When combined with the mass of an object, the density of free gravitons is directly proportional to the planets' orbital speeds. By not inverting the curve as Einstein did, this concept produces every aspect of general relativity but in the mirror image. This new theory explains exactly how the density develops due to pressure. It also functions correctly at the molecular level and explains the speeds of stars around galaxies.

DIAGRAM NO. 9

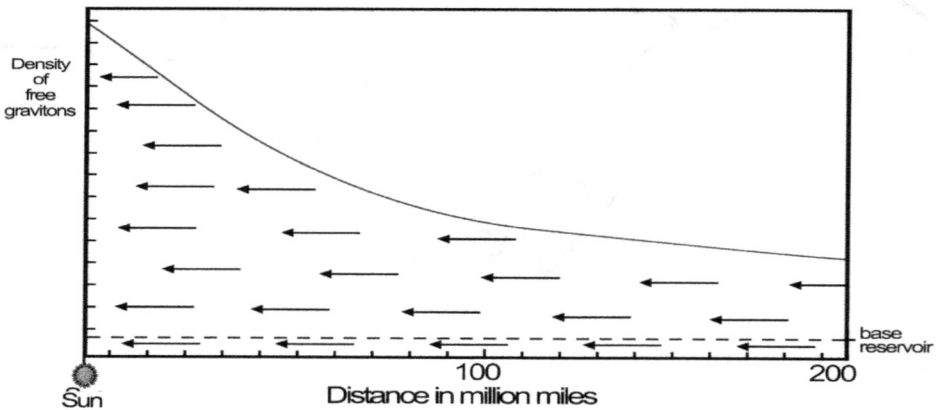

Diagram No. 9 is a depiction of the build-up of free gravitons that have been displaced from the Sun. The bottom of the graph shows the base reservoir. The polarizing of the displaced free gravitons also polarized the base reservoir which is shown by the arrows within the graph. The density of the displaced free gravitons is added onto the base reservoir and produces a curve that is a duplicate of the curve shown in Diagram No. 5. The displaced free graviton density added to the base density along with the polarization produces the amount of gravitational attraction available relative to distance.

DIAGRAM NO. 10

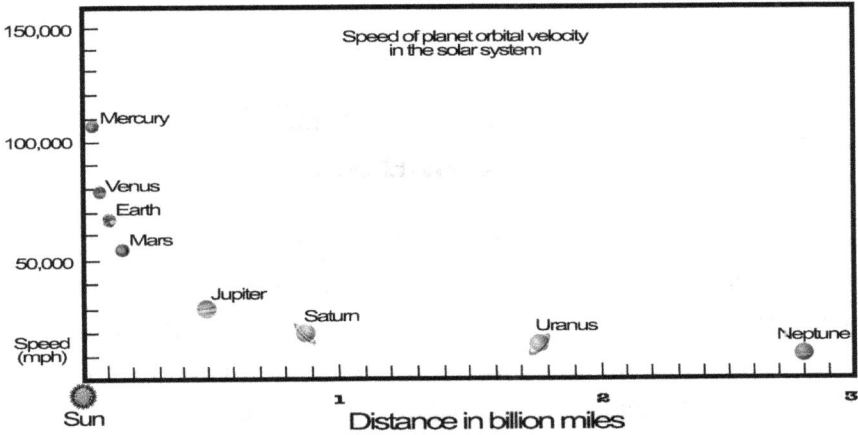

Speed of planet orbital velocity in the solar system

Diagram No. 10 shows the velocity of the planets' orbits on a scale that includes all 8 planets.

DIAGRAM NO.11

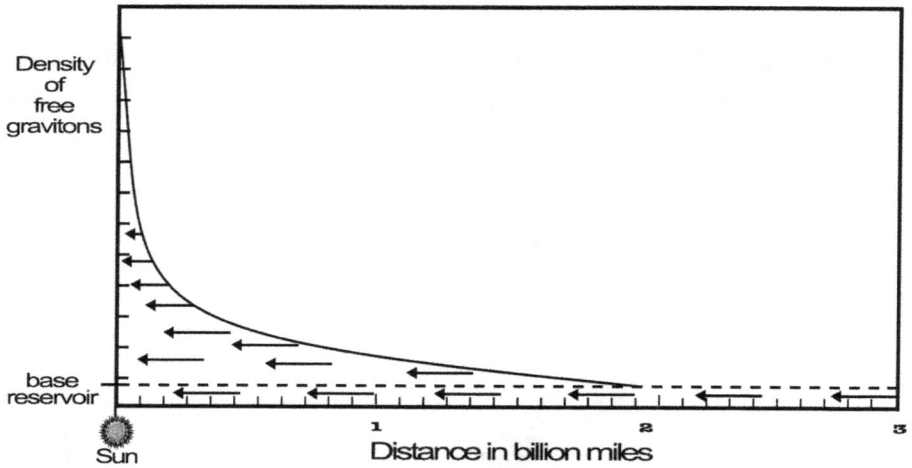

Diagram No. 11 is a depiction of the directly proportional build-up of free graviton density created by the Sun. The alteration in the scale of distance makes the curve appear different from the curve in Diagram No. 9, but it is the same information scaled-up in distance to include all 8 planets. The increased localized density dissipates within this graph but the base density of the reservoir continues throughout the galaxy.

DIAGRAM NO. 12

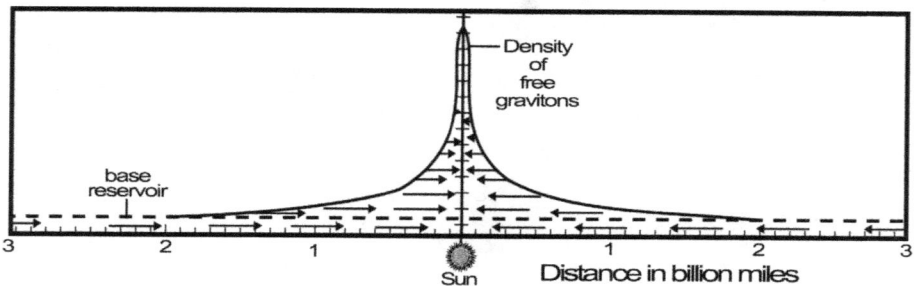

Diagram No. 12 depicts the density of free gravitons displaced by the Sun. They are added onto the base reservoir of free gravitons in

our vicinity of the galaxy. The base reservoir is traveling around the Milky Way at approximately 500,000 mph. The Sun's speed matches this speed because its dominance has weaved itself into the moving reservoir.

The planets each build their own area of polarized increased density that is not shown on this graph, but was shown in Diagram No. 1. The speed of each planet is determined by the quantity of mass interacting with the density of free gravitons it is traveling through. The attraction between the Sun and each planet is determined by the interaction of each mass's field of free gravitons. These two aspects make up the balancing mechanism that matches the speed of each orbit with the amount of attraction between masses.

Diagrams 9-12 used the same information that was clearly the origins of general relativity and placed it into a medium of what is known as dark matter. Along with a hypothesis consisting of minimal observed logical attributes explaining dark matter as free gravitons, I have represented all the outcomes of general relativity. This new concept also explains in detail how the attractive force responds to pressure within mass while clarifying the motion of stars. All distance measurements contained were converted to miles to frustrate the scientific community who snubs any thoughts that are not 100% compliant with general relativity.

MERCURY'S ORBIT

Anyone who looks back at Albert Einstein's origins of general relativity must review the orbit of Mercury. At the beginning of the 20th century the orbit of Mercury did not match the Newtonian calculations of the day. With all the estimates of mass and the calculations of attraction and speed, there was something missing that was perturbing or altering the orbit. Multiple theories were presented but were found not to be correct.

In 1915 the presentation of general relativity solved this problem and also was correct for the orbits of all of the planets. During Mercury's elliptical orbit, the planet was moving through areas of space with differing curvature of space-time. Mercury is thought to be traveling a shorter distance around the curvature closer to the Sun and a longer distance around the curvature when further from the Sun. If all calculations of speed and distance were correct, this was an unbelievable yet possible solution. The prediction that light would follow this same curvature that was confirmed in 1919 cemented general relativity as a known entity instead of being simply a theory. Under my theory of mass displacing free gravitons and holding them close due to their polarization, Mercury's elliptical orbit passes through areas of varying free graviton density. These variations in density result in unforeseen attraction and speed variations that were not accounted for in Newton's theory. The increased speed and attraction at perihelion causes the planet to achieve breakaway velocity later in its orbit. This produces the same orbital results as general relativity without taking a shortcut through space. It is the additional attraction of dark matter that is perturbing Mercury's orbit and altering it from Newton's calculations. Everything about dark matter is the mirror image of general relativity due to Einstein inverting the graph containing the planets' speeds as shown in Diagram No. 4. Mass along with pressure and its motion manipulates dark matter; it does not curve space-time. These differences, when extended to an extremely massive object, also produce the mirror image of what general relativity produces. The extreme curvature resulting in a black hole is instead an extremely dense concentration of free gravitons surrounding a gravitational dark star (GDS) that blocks all light from passing.

Messenger Probe spent over four Earth years orbiting Mercury. The data collected by NASA's radar should be used to plot Mercury's exact speeds and orbital positions to see if Mercury does or does not take a short cut through space.

There is a consistent point at perihelion when Mercury stops rotating on its axis for more than four earth days due to tidal reaction from the Sun. It starts rotating again after Mercury leaves perihelion and moves toward aphelion. To understand this, you must remove the orbital contribution of one rotation per orbit. This changes Mercury's rotation on its axis from 3:2 to 1:2, or one rotation per two orbits.

This is significant because the locking of Mercury in rotation seems to incrementally change its speed. This is why relativity's calculations for Mercury's orbit are off by a small fraction. It also appears that tidal force from the Sun is altering internal pressure on Mercury. This is contrary to current thinking that 100% of all rotation of planets on their axes comes from the conservation of angular momentum during formation.

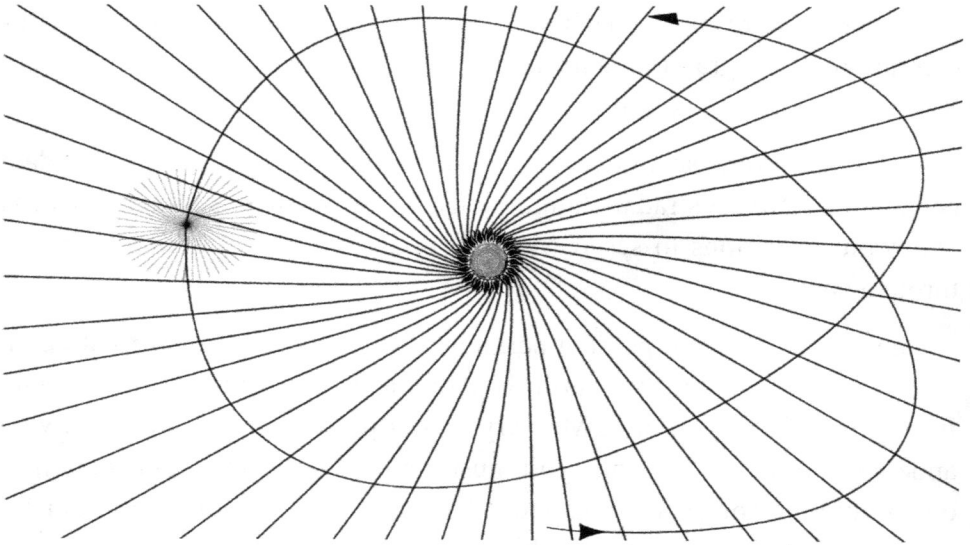

Diagram No. 13 depicts Mercury's elliptical orbit around the Sun. The daisy configuration formed by its elliptical trail is the result of both general relativity's curvature of space-time and the variations in free graviton density. The results of these two theories are identical. Mercury and its localized field of free gravitons is shown at the location where it stops rotating on its axis due to tidal influences from the sun.

LENSING

Mass does not curve space-time. It alters the density and polarizes the reservoir of free gravitons that surround it. The rotation of mass produces a curvature of the free gravitons adjacent to it. The photons and other electromagnetic particles follow the influence of this curvature as they pass through this area.

The fact that Einstein predicted this lensing that was later confirmed many times is in my opinion the biggest coincidence in science. His forecast of light curvature was correct but for the wrong reason. This one prediction was the confirmation of his theory and is why physicists cannot break away from his explanation of its cause. It is also the reason the original parameters of general relativity have been stretched so far and its prerequisite has been ignored. Under my theory of free gravitons, this phenomenon of lensing can be explained but would not be readily forecasted.

DIAGRAM NO. 14

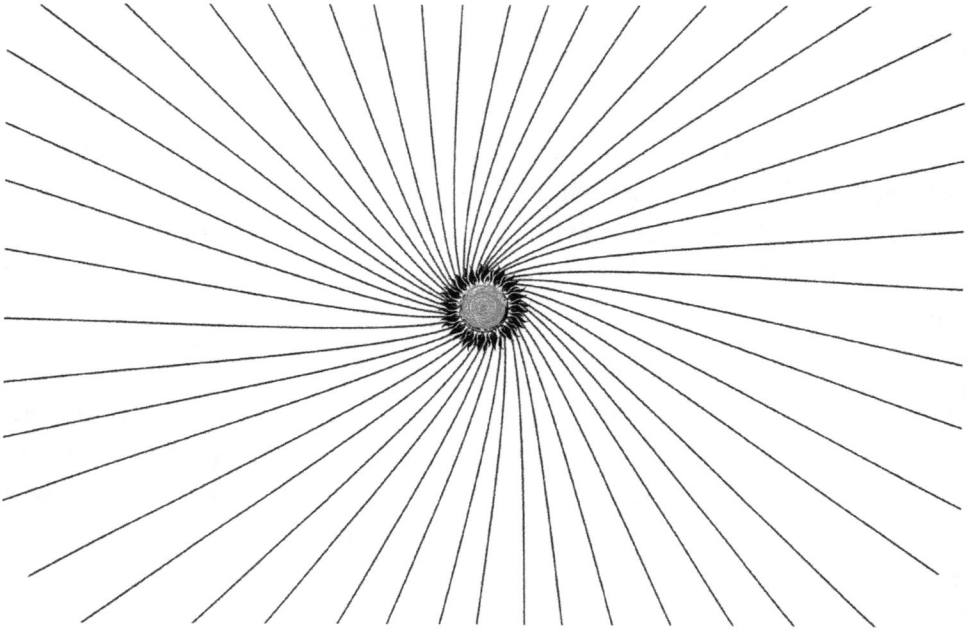

Diagram No. 14 depicts the polarized free gravitons that become denser in close vicinity of a massive object and also follow its rotation. Any light passing through this area from beyond the mass will follow this curvature. Gravitational slip will occur as these connections dissipate beyond the horizon and new connections form near the opposite horizon during rotation.

GRAVITATIONAL WAVES

Early in 2016, a group of researchers announced the detection of gravitational waves at two LIGO facilities. This is good news for anyone attempting to reaffirm general relativity. Obviously, I would question whether researchers who have designed and operated equipment for this purpose can distinguish the difference between a wave in space and a wave in dark matter. This goes back to my original argument: Is space empty or is there a medium of dark matter?

Apparently, they have truly found something and I hope they will consider all options as they move forward. I assume these researchers refer to the base reservoir of free gravitons as dark matter. The term dark matter was coined to literally mean something that is not understood.

Just as with gravitational lensing and the connections gravitational attraction has to time, the existence of gravitational waves can be explained. We now have access to the information that dark matter does exist. Science cannot continue ignoring conflicts with general relativity and only highlight confirmations of it. This produces an incomplete picture of the entire process of how gravitational attraction operates.

It is clear that most if not all of today's gravitational researchers conform to the position that all gravitational theories must pass the general relativity litmus test. Newtonian theory was discarded because it failed this litmus test, and dark matter is not understood as a consequence. Researchers need to be aware there is an alternate explanation for the observed abnormalities such as lensing, etc. and understand that all gravitational theories do not have to conform to the litmus test. Isaac Newton was correct when referring to gravitational force and gravitational fields because that is the contribution of dark matter.

CORIOLIS EFFECT

Atmosphere Related

There is an influence of the gravitational force that is apparent within our upper atmosphere. According to today's science, the rotation of the Earth along with the conservation of angular momentum is 100% responsible for the high and low pressure systems on Earth rotating inversely on opposite sides of the equator. I contend that gravitational attraction is exerting an unseen influence while attempting to hold the atmosphere in place. Free gravitons fill every void around the atoms of mass particularly within the upper atmosphere. These particles are not visible and under general relativity do not exist, yet these particles are present and they direct the movement of air that is rising or falling through their influences. This makes the Coriolis Effect within the atmosphere a side effect of gravitational force and an opportunity to detect free gravitons on Earth. Much stronger forces are being applied to the atmosphere creating motion such as heat and its related winds along with some contribution by the Earth's rotation. These outside forces exerted by rotation might explain low pressure systems nearer the equator and maybe high pressure systems further from the equator.

There are unexplained arctic lows that form quite often and are a contradiction to the predictions of the current Coriolis Effect. All of the outside additions of energy and motion have one thing in common, providing the "chicken or the egg" dilemma. Does the rising or falling of air cause it to rotate in a certain direction? Or does the moving air in a certain direction cause it to rise into low pressure systems or fall into high pressure systems? Usually the first option is true. The rising of warm air or falling of cold air determines the direction of a weather system's rotation. In the case of arctic highs and lows in my theory, however, the latter option is true. In this

40

instance, the direction of rotation determines whether the air rises resulting in a low pressure system or falls becoming a high pressure system. Arctic lows can be extremely cold and still have lower atmospheric pressure due to their direction of rotation. The gravitational influence does not create motion in the atmosphere, but by attempting to hold it in place gravity guides the motion of high and low pressure systems as they rotate. These actions are inverted on opposite sides of the equator but not within 5 degrees of it.

The pressurized churning of the Earth's core appears to be the cause for this influence and its reversal from northern to southern hemispheres. This pressurized churning needs to be credited with initiating spherical masses to rotate on their axes. This also explains how Jupiter's red spot has been a stationary storm for hundreds of years. It is likely a gravitational anomaly created by a hotspot on Jupiter's surface. If you follow this gravitational influence into space, it also limits the formation of rings around a planet to the area around the equator of any planet.

A CONDENSED VERSION OF HOW GRAVITY WORKS

1. Gravity is a force produced and conducted by dark matter, which is a free graviton, that is only visible when it is dense enough to effect the passage of light.

2. Each galaxy is engulfed within a moving reservoir of free gravitons that is following the rotating object at the galaxy's center.

3. Every atom in space commandeers an amount of free gravitons that is proportional to the quantity of gravitons within that atom.

4. The natural balance of gravitons inside and free gravitons clinging to the outside of matter is the foundation of gravity because it enhances the ability of every atom to attract other atoms. This is how matter attracts matter.

5. When two objects each affect the same free gravitons they then attract each other.

6. As mass accumulates its buildup begins to create internal pressure. This internal pressure reduces the space available for free gravitons and causes them to migrate outside that mass. These free gravitons are held close to that mass increasing the density of the reservoir of free gravitons around that object. This is what Isaac Newton was searching for: It is how mass attracts mass.

7.   These free gravitons that have been displaced from their natural position of balance are polarized back toward the mass they have been forced away from. This polarization extends outward through the reservoir many times further than the increased density reaches.

8.   At the point the mass can form itself into a sphere the internal pressure appears to exert a rotational influence on that sphere.

9.   When one object becomes considerably larger than all other objects in its vicinity its dominance takes over that entire area of the reservoir. This establishes the motion of all stars and near stars matching the speed of the reservoir through space.

10.   The entire remaining cloud of particles collapses and rotates around this dominant mass mainly due to the conservation of angular momentum. These remaining objects are left to build their own smaller polarized area of influence that moves through the dominant masses field. (This is the motion that all previous gravitational calculations have been based upon.)

11.   As these remaining objects travel through the polarized reservoir of the dominant mass, any build-up of free gravitons in front of that object acts as an attractive force influencing that object forward instead of being a resistance impeding its motion.

12.   The increase of speed in areas of denser free gravitons or resulting decrease of speed in areas of less dense free gravitons acts as a balance against the variations of attraction due to changes in distance.

13.   In close proximity to a large rotating mass the high density of free gravitons is influenced to follow this rotation due to the polarization which causes a curvature in the free gravitons. Any light passing

through this area will follow this curvature. Mass and pressure manipulate free gravitons; mass does not manipulate space-time.

14.   If a star achieves the correct mass under sufficient pressure, the resulting density of free gravitons surrounding it will completely block light. At this point the star becomes a gravitational dark star (GDS) not a hole in space-time.

15.   The density of free gravitons affects the rate of change and motion. Everything from the ticking clock down to the very vibration of atoms themselves, gravity varies multiple aspects of how we perceive time.

16.   It is time to return to Newtonian Gravitational Theory with gravitational fields and gravitational attraction.

SUMMARY

Would Albert Einstein's conclusions have been the same if he would have had access to all the theories and observations that have been made since his time? It takes someone who is not 100% committed to general relativity to point out that dark matter's contributions produce effects that are the mirror image of general relativity. Blending this with quantum theory and Newtonian theory shows how dark matter makes gravity work. My combining of theories corrects problems at the molecular level and the outer reaches of our solar system. It also solves the orbital problems of galaxies' discs and produces a gravitational dark star that does not lose any information from the universe. Diagram No. 2 clarifies the motion of stars within galaxies' discs as being a completely separate type of motion. Diagrams 11 and 12 show how the actual speed and distance of the planets in our Solar System are directly proportional to the density of free gravitons. The base reservoir of free gravitons expands the attraction of our Sun to influence a much larger area.

It is intuitively obvious to me that the graviton occupies the space surrounding the nucleus of matter. The free gravitons that make up what is now known as dark matter surely duplicate the actions of the graviton but at a lower density. My vision of a graviton may not match exactly what scientists currently consider it to be, but there seems to be room for discovery in this area. My theory may not solve all the problems with other gravitational theories, but it most certainly is better than avoiding the conflicts with general relativity.

Thanks to Rob Latta for her patience and assistance in editing this book. Thank you to my loving wife Roseann for her support. To my daughter, Jenn Lederer, for helping me with her organizational contribution. Thank you to Nik Le Sante for the graphs and sketches. For additional support I would like to thank Walter Day, Ed Kearns, and Lucas Kearns.